U0166507

双语版·全4册

HEDGEHOG

亲亲动物

如果刺猬来我家

[英]弗朗西斯·罗杰斯 [英]本·克里斯戴尔 著绘　范晓星 译　朱朝东 丁亮 审校

中信出版集团 | 北京

图书在版编目（CIP）数据

如果刺猬来我家：汉文、英文／(英) 弗朗西斯·
罗杰斯, (英) 本·克里斯戴尔著绘；范晓星译. -- 北
京：中信出版社, 2023.3
（DK亲亲动物：双语版：全4册）
ISBN 978-7-5217-5239-7

I. ①如… Ⅱ. ①弗… ②本… ③范… Ⅲ. ①猬科—
少儿读物—汉、英 Ⅳ. ①Q959.831-49

中国国家版本馆CIP数据核字（2023）第021859号

Original Title: How can I help Roly the hedgehog?
Copyright © 2022 Dorling Kindersley Limited
A Penguin Random House Company
Simplified Chinese translation copyright © 2023 by CITIC Press Corporation
All Rights Reserved.

本书仅限中国大陆地区发行销售

致所有好奇的孩子！

如果刺猬来我家
（DK 亲亲动物双语版　全 4 册）

著　绘：[英] 弗朗西斯·罗杰斯　[英] 本·克里斯戴尔
译　者：范晓星
出版发行：中信出版集团股份有限公司
　　　　　（北京市朝阳区东三环北路 27 号嘉铭中心　邮编　100020）
承　印　者：北京顶佳世纪印刷有限公司

开　　本：889mm×1194mm　1/20　　　　印　张：2　　字　　数：115 千字
版　　次：2023 年 3 月第 1 版　　　　　印　　次：2023 年 3 月第 1 次印刷
京权图字：01-2022-4478　　　　　　　　审　图　号：GS 京（2022）1525号
书　　号：ISBN 978-7-5217-5239-7
定　　价：156.00 元（全 4 册）

出　　品　中信儿童书店
图书策划　红披风
策划编辑　陈瑜
责任编辑　虔慧
营销编辑　易晓情　李鑫樘　高铭霞
装帧设计　哈_哈

For the curious
www.dk.com

混合产品
纸张｜
支持责任林业
FSC® C018179

版权所有·侵权必究
如有印刷、装订问题，本公司负责调换。
服务热线：400-600-8099
投稿邮箱：author@citicpub.com

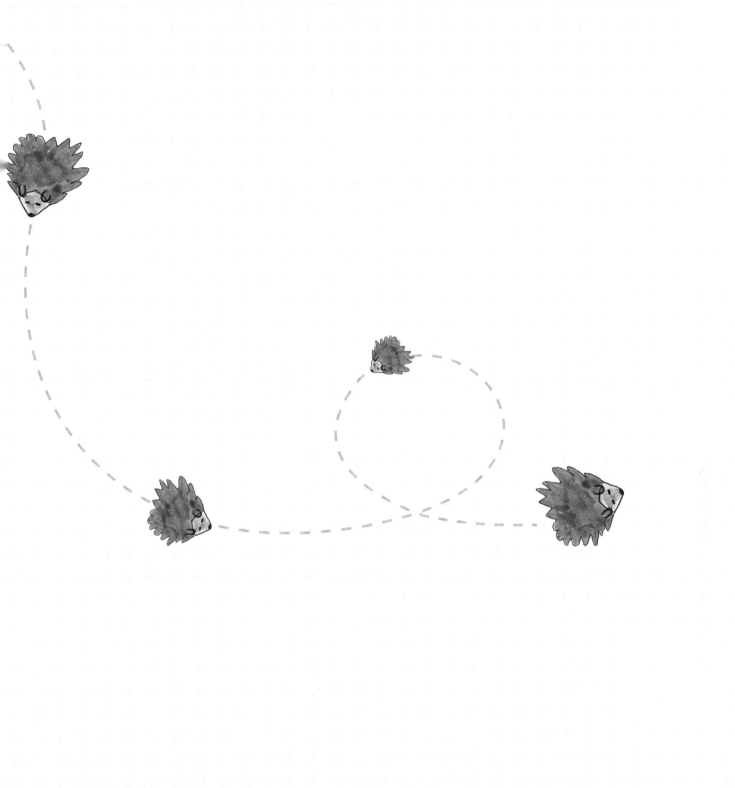

你好，我叫罗里。我是一只小刺猬，
我想去你家花园玩一玩。
可不可以请你帮个忙？

Hello, my name is Roly.

I am a hedgehog and I like to visit your garden.

But I need your help.

我怎么才能进到你家花园呢？
先请大人在木栅栏上
给我做一个小门吧。

Please let me into your garden.

Ask a grownup to make me a doorway in your fence.

12 厘米（12cm）

你家花园里有
我喜欢吃的东西。

I like to eat things that I can
find in your garden.

请你种一些花花草草，
这样鼻涕虫和昆虫就会来啦。

Please plant flowers to
attract the slugs and bugs.

我口渴了。

I get very thirsty.

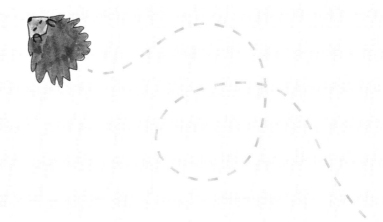

请给我留一小碗清水。

Please leave a small
bowl of water out for me.

我有可能被垃圾卡住。

I can get stuck in rubbish.

请你把花园收拾得干净整洁吧。

Please keep your garden tidy.

垃圾桶
Bin

这里有池塘，我可要小心！

Ponds can be a danger to me.

请给我留一个梯子吧，
万一我掉进水里，就可以爬出来啦。

Please give me a ladder
in case I fall in.

我还有可能被球网困住。

I can also get stuck in nets.

不用球网的时候，请把它收起来吧。

Please lift them up

when not in use.

我爱在树叶堆和肥料堆里睡觉。

I sleep in piles of
leaves and compost.

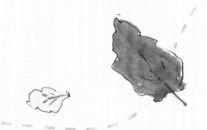

所以当你清理树叶和肥料的时候，
要小心看一下，我有没有在里面哟。

Please be careful and look for me
before you clear them up.

谢谢你为了保护我们做的一切。

Thank you for all your help.

我们为什么要保护刺猬？

Why do we need to protect hedgehogs?

我们要保护像罗里这样的小刺猬。刺猬主要生活在欧洲、亚洲、非洲和大洋洲的新西兰，但其实世界上的刺猬已经不是很多了。

Hedgehogs like Roly need to be protected. Hedgehogs are found in Europe, Asia, Africa, and New Zealand but there are not many of them left in the world.

科学家担心这些可爱的小动物很快就会灭绝，也就是说，在这个地球上，我们再也看不到它们了。因为它们找不到适合睡觉的地方，也经常找不到足够的食物。

Scientists are worried that these little animals will go extinct soon, which means that they won't exist any more, because they are struggling to find places to sleep and enough food to eat.

小刺猬非常需要帮助，就看我们的了！
It is up to us to do what we can to help!

会保护自己的"小刺球"
Prickly protection

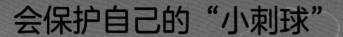

棘刺
Spines

刚出生的刺猬宝宝有个可爱的英文名字，叫 hoglet。它们出生时，皮肤皱皱的、粉粉的，上面覆盖着一些软软的、白色的棘刺。等到它们慢慢长大，棘刺会变得越来越锋利、坚硬，也越来越长。

Newborn hedgehogs are called hoglets. When they are born, hoglets have a few soft, white spines covering their wrinkly, pink bodies. As they grow up, their spines become sharper, harder, and longer.

刺猬宝宝
Hoglet

完全成年的刺猬身上能有多达 7000 根棘刺！
Fully grown hedgehogs can have as many as 7,000 spines!

刺猬是弱小的动物，所以当它们遇到危险的时候，身上的棘刺就能派上用场了。它们害怕或者睡觉时，会缩成一个"小刺球"。

These spines can come in handy when these small animals are in danger. When they get scared or when they are sleeping, they curl up into a prickly ball.

当它们的身子蜷起来时，看起来像个松塔。这样它们身上柔软的部分，比如尾巴、脸、肚子和腿就会被保护起来。

When they are curled up, hedgehogs look a bit like pine cones. This protects the soft parts of their body, such as their tail, face, belly, and legs.

你在花园里见过"松塔小刺猬"吗？

Can you see any hedgehog pine cones in your garden?

冬眠
Hibernation

刺猬是非常爱打瞌睡的小动物，一生中很多时间都在睡觉。有些刺猬一天能睡 18 小时，在寒冷的冬天还要冬眠好几个月。

Hedgehogs are very dozy creatures who spend a lot of their life asleep. Some hedgehogs can sleep for up to 18 hours a day and need to hibernate (rest) through the chilly winter months.

一些动物，比如刺猬，会在冬天进入深度睡眠，这叫冬眠。这是因为冬天天气寒冷，很难寻找到食物。不过在开始冬眠之前，刺猬要去收集很多食物。

Hibernation is when animals, including some hedgehogs, go into a deep sleep because the weather is too cold and it is hard to find food. But before they hibernate they must gather lots of food.

毛毛虫
Caterpillars

蠼螋
Earwigs

遗憾的是，这些浑身是棘刺的小家伙视力不是很好，所以它们利用听觉和长长的鼻子去寻找食物，比如鼻涕虫、毛毛虫。

Unfortunately, these spiny creatures don't have very good eyesight so they have to use their hearing and their long snout to hunt creatures such as slugs and caterpillars.

蚯蚓
Earthworms

鼻涕虫
Slugs

马陆（千足虫）
Millipedes

刺猬寻找食物的时候，可以走 3000 米那么远，对这些短腿的小家伙来说，这是很远的路程！

When they are out looking for food, hedgehogs can walk for up to 2 miles (3 km) – that's a long way for little legs!

好吃！
Yum!

鸣 谢

本出版社感谢以下机构提供照片使用权：

(a= 上方；b= 下方；c= 中间；f= 底图；l= 左侧；r= 右侧；t= 顶端)

34 Dreamstime.com: Cynoclub (cl); Eric Isselee (cr). **35 Dreamstime.com:** Eric Isselee (bl). **39 Dorling Kindersley:** Natural History Museum, London (clb). **Getty Images:** Photographer's Choice RF / Jon Boyes.

其余图片版权归英国 DK 公司所有，更多信息请访问 www.dkimages.com。

关于作者和绘者

本和弗朗西斯是一对夫妻，住在英国泰恩河畔纽卡斯尔。他们倾心于帮助家中花园里的野生动物。仲夏的夜晚，弗朗西斯醒来，有了一个灵感——创作鼓励小朋友加入他们的行动的系列童书。

弗朗西斯创作故事，本画插图，他们引领小读者进入一个栩栩如生的野生动物世界，快来欢迎花园里的小客人：小刺猬罗里、小麻雀罗芮、小蜜蜂罗丝、小蝴蝶罗克西。